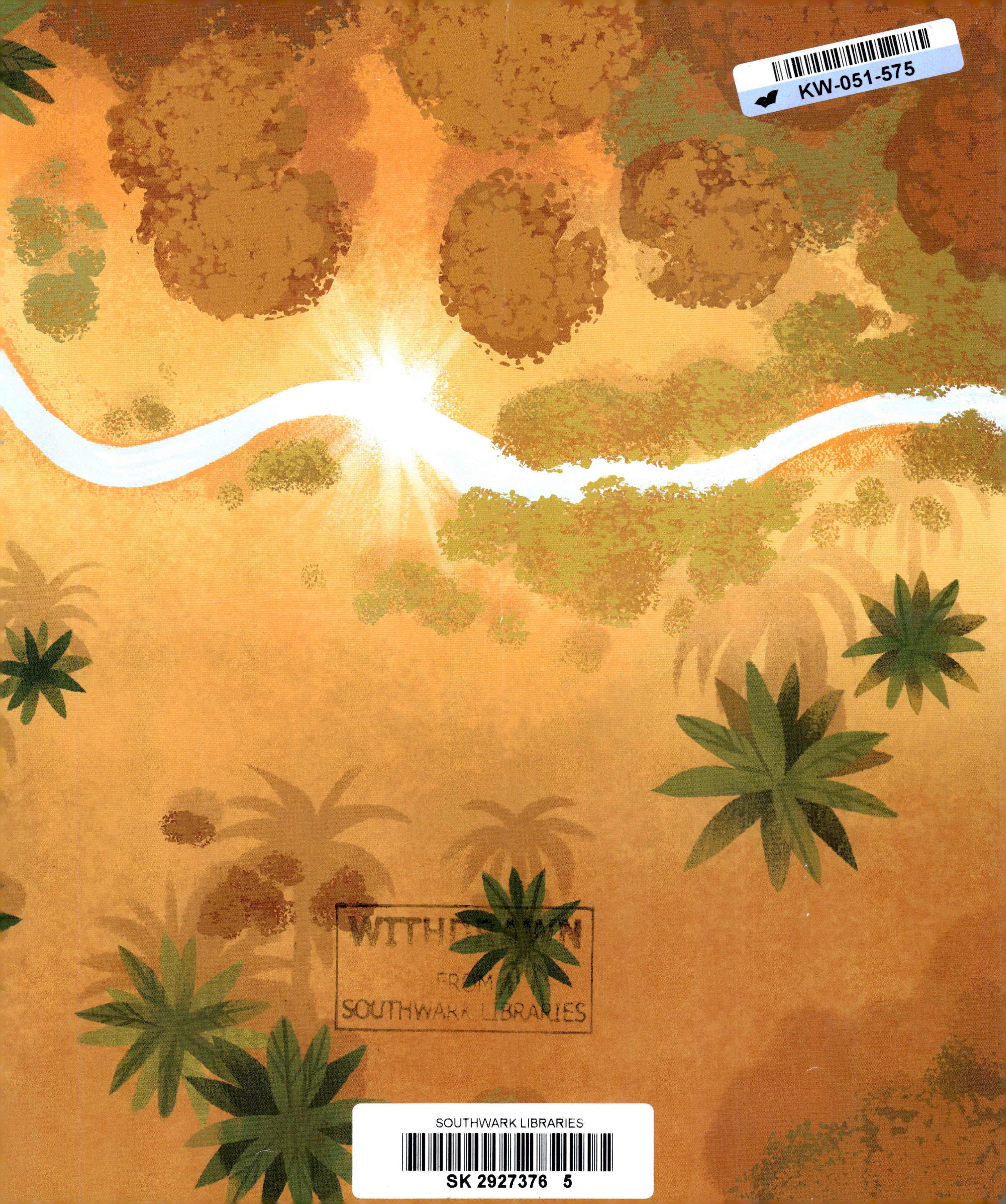

The big and the small, the fast and the slow, the strong and the weak, the fierce and the timid, the weird and the wonderful – it takes all sorts to make our wild world work. And sometimes it disguises success as failure: when giant trees fall, new saplings grow; when enormous whales die, creatures on the deep ocean floor feast; when at last a species vanishes through extinction, there is a gap for something new to evolve. And that is the greatest wonder of all: understanding how all of this life, how all of our planet, works together.

We will never know all the questions, let alone the answers, but each time we peer a little closer at the natural world, we uncover remarkable stories. Some are told far away, some unravel in our own backyards, some are surprising – almost unbelievable – and some are so simple we should have known them all along.

For me, science is the art of understanding the truth behind nature's beauty. And here, woven through the fabulous art, you'll find the truly amazing stories of beautiful life across our planet. A collection of exciting characters coming together to present the greatest show the Earth has ever seen. *Wow*!

But they are all in danger and we are in danger too. The cast of life is rich, strong and defiant – nothing wants to die. But we are tearing down the stage, burning the forests, polluting the seas, poisoning the land and choking the air. We know we must stop, we know how to heal the Earth, but we need your help and the best fuel for change is . . . love. A love for everything that creeps, crawls, slithers, slimes and stings: a love for life, all life. So when you read and look at this book, please see it as a love letter from Planet Earth, sealed with the kisses of millions of beautiful living things.

– CHRIS PACKHAM

PLANET EARTH III

PUFFIN BOOKS

UK | USA | Canada | Ireland | Australia | India | New Zealand | South Africa

Puffin Books is part of the Penguin Random House group of companies
whose addresses can be found at global.penguinrandomhouse.com.

www.penguin.co.uk   www.puffin.co.uk   www.ladybird.co.uk

Penguin
Random House
UK

First published 2023
001

Written by Leisa Stewart-Sharpe
Text copyright © Children's Character Books Ltd, 2023
Illustration copyright © Kim Smith, 2023
Introduction copyright © Chris Packham, 2023
BBC and BBC Earth (word marks and logos) are trade marks of the
British Broadcasting Corporation and are used under licence
BBC logo © 1996, BBC Earth logo © 2014

The moral right of the illustrator has been asserted
Produced in consultation with the Planet Earth III production team at BBC Studios Natural History Unit

Printed in Italy

The authorized representative in the EEA is Penguin Random House Ireland,
Morrison Chambers, 32 Nassau Street, Dublin D02 YH68

A CIP catalogue record for this book is available from the British Library
ISBN: 978-1-405-94670-4

All correspondence to:
Puffin Books, Penguin Random House Children's
One Embassy Gardens, 8 Viaduct Gardens
London SW11 7BW

FSC
www.fsc.org
MIX
Paper | Supporting
responsible forestry
FSC® C018179

BBC

PLANET EARTH III

LEISA STEWART-SHARPE and KIM SMITH

PUFFIN

A
CONNECTED
PLANET

Planet Earth is an extraordinary place. It's a place of tangled forests, rolling grasslands and sun-baked deserts. Of freshwater worlds, expansive oceans and coastlines carved by the waves. It's extreme in almost every way – extremely hot, cold, dark, high and *beautiful*.

But above all, Earth is HOME to countless plants and animals and eight billion humans, all depending on each other for survival. And with each passing day we discover just how deep those connections go.

Imagine a beautiful web of life, woven from invisible threads, that connects every living thing to each other. Energy flows through each thread, from the tiniest algae to the mightiest whale, so that ecosystems can exist in perfect balance. Humans are very much part of this web of life too, but as our numbers have grown, we've begun to upset nature's balance even in the furthest reaches of the planet. From farming and mining to fishing, we've created clever, but sometimes damaging, ways to take what we need from Earth. It's clear we're taking too much. And now, our most important lesson has begun – learning to take care of the planet, so that it can take care of us.

Come on a journey to the wildest corners of Earth and meet some of the incredible wildlife we share it with. Learn about the fragile connections forged in nature over millions of years, and how each of us can help protect them for a million more.

It's time to experience the beauty and majesty of Planet Earth.

THE TREE OF LIFE

Within one of Earth's oldest jungles stands one of its tallest trees – the KAPOK.

She is a giant in the forest. To some of the Indigenous Peoples who share her Amazon rainforest home, the kapok is known as 'vovo', or grandmother. For she understands that in the crowded forest, survival is about taking care of each other. If the world around her thrives, so will she. And so, the kapok is food, she is home, she is a bustling city in the sky for life throughout the forest.

The kapok's umbrella-shaped crown rises above the canopy, some 70 metres high. Up here above the jungle, HOWLER MONKEYS *whoop*, and HARPY EAGLES *swoop!*

HARPY EAGLE

SCARLET MACAW PARROT

BAT

BARE-TAILED WOOLLY OPOSSUM

HOWLER MONKEY

SPIDER MONKEY

SPIDER MONKEYS hang from the kapok's giant branches.

A pandemonium of **PARROTS** creates a racket as they flock to the kapok's flowers. As night falls, the flowers attract **BATS**. After the flowers come the fruits, and when they burst open, they look like cotton wool inside. The fibres blow far and wide, scattering the kapok's seeds across the forest.

There are worlds here within worlds: a **BROMELIAD** *(brow-mee-lee-ad)* plant clings to the kapok – this cup-like plant creates a pool for a **FRINGED LEAF FROG**.

TROPICAL THORNYTAIL IGUANAS scramble downwwwwwwwwwwwwwwwwwn the kapok's trunk.

BROMELIAD

FRINGED LEAF FROG

TROPICAL THORNYTAIL IGUANA

EPIPHYTE

The kapok is a hanging garden, with **EPIPHYTE** (pronounced *eh-puh-fite)* plants growing off her branches and dangling in the air.

TOUCAN

It's a well-travelled highway to the kapok's buttressed feet, where wide roots support her great weight. Down here, thousands of insect species *creep* and *crawl* across the dark forest floor.

This grandmother tree is a beautiful reminder of the way in which all living things are connected, sometimes in a habitat no bigger than a single tree.

LEAF-CUTTER ANT

FORESTS
FRIENDSHIPS AMONG THE LEAVES

Most life on land is hiding in the deepest reaches
of our forests, with many species still unknown.

From uncovering incredible animal behaviours inside Earth's seasonal temperate
forests, to peering inside microworlds ruled by mini-monsters in Earth's steamy tropical
rainforests, scientists make thousands of beautiful and bizarre discoveries every year.

Zoom in on the crowded Ecuadorian rainforest, and on a tiny stem
is the even tinier world of the **TREEHOPPER**. About as small as a
corn kernel, these sap-sucking insects scurry about in *terrifying*
disguise, wearing *weird* helmets, to fool pesky predators.

And if you thought their helmets were outrageous,
if only you could hear their song. Although
inaudible to humans, treehoppers sing by sending
vibrations from their abdomen, down their limbs
and through the plants they live on.

TREEHOPPER

It's their own secret language. These babies are using it now to raise the alarm.

ANT

Brrrrr bup brrrrr. They're saying: 'Help, we're under attack!'

ANTS are getting too close.

But with a swift SWAT, CHOP and KAPOW, the babies' mum sends them tumbling.

But when a much bigger **ASSASSIN BUG** appears, Mum realizes she's going to need help. But who?

ASSASSIN BUG

STINGLESS BEE

Coming in loud and fast are **BEES**, swiftly scaring off the assassin bug. But *uh oh*, instead of flying away, the bees *land* on the babies' backs. And in a strange twist, they begin . . . to tickle.

The grateful babies respond, rewarding their bodyguards with their delicious, sugary poos.

Complex connections such as this have been forged over millions of years to the mutual benefit of so many living things – whether between plants and pollinators, or a parasite and its host.

Journey into the planet's forests and discover how the actions of each plant and animal keep Earth's ecosystems in perfect balance.

STORIES
FROM THE FORESTS

SPIRIT OF THE FOREST

When you think of rainforests, you might imagine steamy jungles drenched in rain, but in the north-west of the American continent, there are milder temperate rainforests that can be blanketed in snow.

These forests are habitats of enormous proportions; old giant fir and spruce trees stretch up, as furry giants roam below. And hidden here is one of the **rarest** of all bears.

In western Canada's Great Bear Rainforest, white fur shimmers in the sunlight as though a ghost is moving through the forest.

This **KERMODE BEAR** is special. Unlike most of her kind, who are black, this bear is completely white, from the tips of her ears right down to her claws.

She is known as a spirit bear – it's thought less than one hundred and fifty of these white bears remain in the wild.

At over twenty years old, this grandmother is perhaps the oldest of them all.

She spends most of the year concealed within the forest, but today she heads for the river.

The salmon are running!

Each autumn, an impressive migration passes through this forest. The **PACIFIC OCEAN SALMON** make a difficult journey, swimming from the sea and struggling up the rivers into creeks to lay their eggs.

And as they do, the carnivores are waiting. There are bears *everywhere*.

She takes note, lunges, grasps and

S P L A S H !

The old spirit bear *swipes* . . . and misses. She isn't as sprightly as she used to be – or perhaps the salmon are getting more slippery? She could learn a thing or two from the young **BLACK BEAR** crashing through the water beside her.

Time to return to the forest to eat in peace.

Like the other bears, this grandmother leaves the fish carcass behind when she's done, so the real magic can begin. As it decays, it's absorbed into the forest floor, releasing nitrogen that fertilizes the forest.

Without the fish, the bears wouldn't eat. And without the bears, the forest wouldn't flourish.

They're all connected in a beautiful, yet fragile, web of life.

INHABITANTS
OF THE FORESTS

SNEAKING UP ON LOVE

In the misty bamboo forests of central and southern China, a type of pheasant, known as the **TEMMINCK'S TRAGOPAN**, is ready for love. However, he's rather shy.

When a female wanders close by, he . . . hides, waiting for the right moment to reveal himself. He clacks and flaps, raises his horns, inflates his blue bib and . . . PEEKABOO! What an entrance.

MALE TRAGOPAN

FEMALE TRAGOPAN

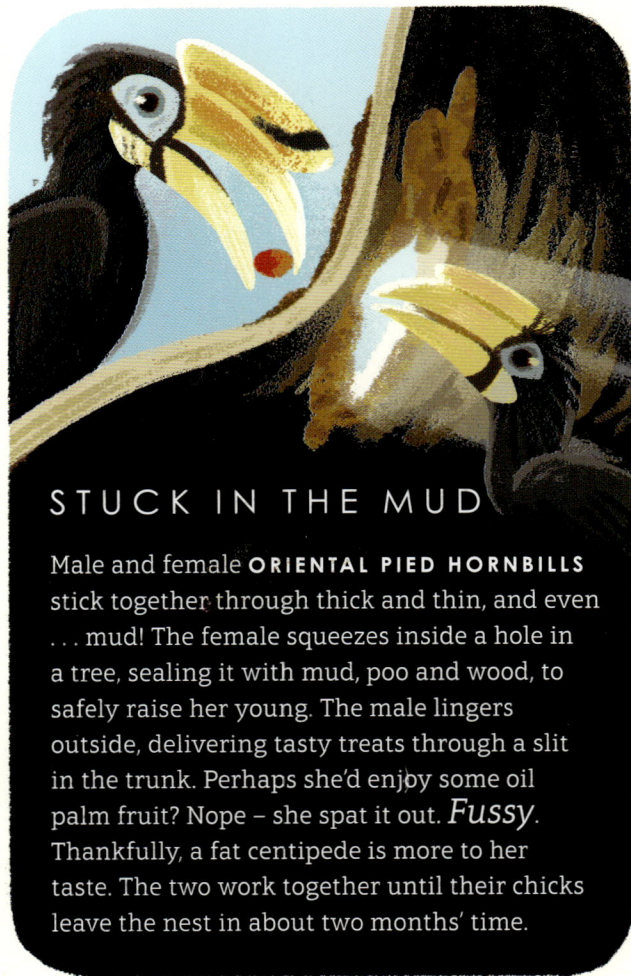

STUCK IN THE MUD

Male and female **ORIENTAL PIED HORNBILLS** stick together through thick and thin, and even . . . mud! The female squeezes inside a hole in a tree, sealing it with mud, poo and wood, to safely raise her young. The male lingers outside, delivering tasty treats through a slit in the trunk. Perhaps she'd enjoy some oil palm fruit? Nope – she spat it out. *Fussy.* Thankfully, a fat centipede is more to her taste. The two work together until their chicks leave the nest in about two months' time.

THESE ARE NOT DEAD LEAVES . . .

. . . they're millions of **MONARCH BUTTERFLIES**. They've spent several weeks on a wing-wearying flight, migrating almost 4,000 kilometres. They come from as far north as Canada, to begin their long winter slumber in Mexico's warmer mountain forests. The forest works like a blanket, keeping out the cold air and rain. Then, when spring arrives, the butterflies wake, the Sun warms their wings and the march of the monarchs begins again as the butterflies fly back north for the summer.

MONARCH BUTTERFLY

AT A
CROSSROADS

It's early morning in a village in western Uganda, Africa, and the children are setting off for school. But they're not alone . . . a family of wild CHIMPANZEES are starting their day too.

They wait and listen. They look left, they look right, and one by one, they all cross the road. (The littlest chimps get a piggyback.)

This unusual sight has become the norm for humans and chimps here, where the two are learning to live side by side. But it has not been easy.

In the past, the villagers cut down the forests surrounding their homes to sell the trees as wood or charcoal, so they could pay to send their children to school. The vast forest was broken up into little islands, each cut off by farms and roads. The chimps became marooned. No longer able to safely reach other patches of the forest to forage for food, they began to raid the humans' farms.

To help avoid conflict between humans and chimps, conservationists now pay for the local children to attend school. In return, the villagers are planting hundreds of thousands of trees in corridors, which help reconnect the forest islands.

These corridors are a route for the chimps, and a path towards both our worlds existing together – in peace.

DESERTS AND GRASSLANDS
WAITING FOR WATER

Deserts and grasslands, which include savannahs and steppes, make up some of Earth's vast and unforgiving plains. These flat and mostly treeless biomes usually exist next to each other, stretching across more than half of Earth's land.

In the dry season, wildlife may seem sparse here, yet extraordinary survivors are patiently waiting for the most precious of all things . . .

rain.

Its power is life changing – deserts spring to life, grasslands bloom and it's just what *everything* has been waiting for.

LION

Sandwiched between the Sahara Desert in the north and rainforests in the south is Zakouma in Chad, Central Africa. It's a place of two seasons – wet and dry. In the wet, Zakouma's grasslands are awash with water. To avoid the floods, **ELEPHANTS**, **GIRAFFES** and **ANTELOPES** migrate. But when it's the dry season, Zakouma is a magnet for life.

KORDOFAN GIRAFFE

Millions of **RED-BILLED QUELEA** (*kwe-lee-uh*) birds flock here, along with herds of antelope, **BUFFALO** and elephants. But they're all keeping a watchful eye out for predators. **LIONS** and **HYENAS** lurk in the grasses, biding their time, as **CROCODILES** move through shallow waters. **VULTURES** and **MARABOU STORKS** take care of any scraps left behind. It's an ecosystem in perfect balance.

So long as there is water, there is life.

SPOTTED HYENA

But across our rapidly warming planet, there are places where the rains fail or the droughts stretch on. As grasslands around the world are over-farmed, over-grazed or over-cleared, the soil becomes stripped of its nutrients so that no amount of rain will encourage plants to regrow. That's when **deserts** can take hold.

Travel across Earth's great plains and witness the extraordinary ways wildlife is holding on.

AFRICAN ELEPHANT

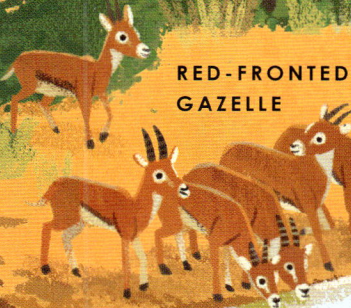

ROAN ANTELOPE

RED-FRONTED GAZELLE

WEST AFRICAN CROCODILE

RED-BILLED QUELEA

BLACK CROWNED CRANE

MARABOU STORK

LELWEL HARTEBEEST

STORIES
FROM THE DESERTS AND GRASSLANDS

SECRETS IN THE SAUSAGE TREE

It's the dry season on Zambia's open plains in southern Africa, and food is scarce.

Fortunately there are flowers. The **SAUSAGE TREE** is blooming; its large red flowers appear to be dangling as if hanging from ropes, providing a welcome breakfast for all manner of hungry beasts from **HIPPOS** to **ANTELOPES**.

But they are also ... A TRAP!

High up in the branches, quietly biding her time, is the most secretive of Africa's big cats – the **LEOPARD**. She's using the sausage tree not only as a hideout, but as bait.

And it's working.

Here comes an **IMPALA** now.

Sshhhhh ...

The leopardess must not give herself away. If she drops too soon, this lone hunter risks injuring herself on the hard, sun-baked ground, and her lunch will get away.

The unwitting impala is almost in position. The leopardess dares not twitch a whisker.

The impala takes another step closer.

Wait for it ...
One more step, and ...

GO!

The leopardess LEAPS on to lunch.

THE WAITING GAME

As the Sun rises over the Tiras Mountains of Namibia in southern Africa, a pair of OSTRICHES wander through the hazy desert heat, leaving the shade of the mountains behind.

Out here it is far too hot and exposed for most of their predators, which makes it the safest place to nest. Just so long as these ostriches can handle the heat.

Mum takes the day shift, shielding the eggs from the **sweltering** Sun.

Dad takes the night shift, guarding the eggs through the bitter cold, while hungry eyes watch him.

So it goes, for around forty days and forty nights, until . . .

CRACK!

The first chick pushes through.

Now it's a waiting game.

The eggs won't hatch all at once, and while these parents wait, the newborn chicks struggle in the sun.

They wait.

The chicks are getting weary now.

They wait for as long as they can . . .

but sadly, time is up.

With the Sun beating down, the family has no choice but to leave. They set off for the safety of the shade, the dry wind blowing at their backs. But wait . . . a faint noise behind them.

CHEEP.
The wind dies down and Dad hears the noise again:
CHEEP.
CHEEP.

It's another chick!

Dad quickly retraces his steps and the little straggler is safe at last.

INHABITANTS
OF THE DESERTS AND GRASSLANDS

SUPER-SNOUTED SAIGA

This unusual-looking fellow is a kind of antelope, known as a **SAIGA**, found in the grasslands of Central Asia and Russia. He has carrot-like horns and a trunk-like nose that helps to filter out any dust he might breathe in. More than a million once roamed the land, but today they're critically endangered.

SAIGA

SHAPESHIFTING DESERTS

Deserts are exposed, ever-changing worlds that are whipped by the wind and scorched by the Sun. Some can shift from bitterly cold to blisteringly hot in the space of a day. As hot air rises from the sand, cooler air swoops in, causing powerful winds to stir, reaching speeds of 100 kilometres an hour.

A LONG TIME BETWEEN DRINKS

On the edge of the Namib desert in Namibia, fifty **CHACMA BABOONS** have formed a *disorderly* queue. Water, thousands of years old, is bubbling up from deep underground, oozing from small holes in the rock. The baboons impatiently wait their turn, as though queueing at a water fountain. When they reach the front of the line, they eagerly press their mouth to the hole and *slurp*.

CHACMA BABOON

SECRETS IN THE CERRADO

Something is hiding in the metre-high grasses of the Cerrado, South America's savannah grasslands.

It looks a little like a fox on stilts, but with the face of a hyena . . .

It's the rarely seen **MANED WOLF**. But unlike any other wolf, maned wolves don't move in packs, instead mostly living alone, even when raising their pups. Yet their secret dens are now being exposed.

Every month vast patches of the Cerrado are cleared to make room for cattle ranches and soy farms. Almost half of the whole savannah has already gone, vanishing at least four times faster than the Amazon rainforest.

The Cerrado is home to 5 per cent of all Earth's biodiversity, including **GIANT ANTEATERS**, **GIANT ARMADILLOS** and **TAPIRS** and absorbs huge amounts of carbon dioxide from the atmosphere that would otherwise make Earth hotter. But it is largely *unknown*. That's why conservationists are fighting to protect the land, a fight we can all support by telling everyone we know about the secrets of the Cerrado.

FRESHWATER

WHERE FRESHWATER FLOWS

JAGUAR

SPIDER MONKEY

Planet Earth is a water world, but it's one of salty seas and ice – water that's either undrinkable or unreachable to most living things. It's *freshwater* that nourishes life across the planet.

Yet unbelievably, only 3 per cent of all Earth's water is fresh and less than 1 per cent of this *flows* as rivers or lakes.

The rest is bound in ice and hidden deep underground.

AMERICAN EEL

Freshwater, which falls from the sky as raindrops, hits the earth and runs down the land. It forms little streams that grow into rivers which feed lakes, spill across plains or cascade down waterfalls. But on Mexico's Yucatán Peninsula, there is barely any freshwater to be seen at all, yet a lush, green rainforest flourishes. So where *does* the freshwater flow?

MEXICAN CROCODILE

CATFISH

The Maya Peoples of Mexico and Central America have always known how to find it . . .

COATI

Even today, they listen for the call and look out for a flash of feathers from the **TURQUOISE-BROWED MOTMOT** bird. It leads them to their water source: the secret underground world of cenotes (sen-oh-tees). These are sinkholes that form when the limestone bedrock on Earth's surface collapses, revealing a secret water world below.

Light from the cenote entrance filters into eerie underground chambers that work like rivers, carrying freshwater from the forest to the sea. Down here, life is in abundance – from fish and turtles to crocodiles and crabs. But dive deeper into these freshwater pools and they soon become dark . . .
and
toxic.

CENOTE MOLLY

CICHLID

TURQUOISE-BROWED MOTMOT

Trapped below some cenotes' freshwater is a misty layer of hydrogen sulphide – a dangerous gas cloud caused by decomposing vegetation.

Even after decades of exploring, cave divers have barely begun to map how far these caves and rivers reach.

Follow the world's freshwater as it flows across the planet, creating opportunities for life to thrive.

STORIES
FROM FRESHWATER WORLDS

THIS IS ONE HOT CROC

During dry season in Yala National Park, on the east coast of Sri Lanka, **MUGGER CROCODILES** wallow in water to cool down.

But in the heat, this mugger's freshwater pool has turned into a puddle. The time has come to haul out.

His short, stubby legs may not look up to the job, but he determinedly sets off, walking many kilometres through thorny scrub until he reaches a large pond.

Alas, it's crowded with other crocs, and these muggers are mega – some adults stretching 5 metres long.

Deciding to play it safe, this smaller male slides in, sticking to the pond's edge. The mud barely covers his back, but he hopes it'll do.

As the Sun beats down, thirsty **CHITAL DEER** leave the forest shade to drink, but the poorly hidden mugger is quickly spotted. He's going to have to do better than that. He sinks deeper into the mud.

He waits.

A deer takes a wary step towards the pond. The **LANGUR MONKEYS** sound the alarm, but the deer is already knee-deep in the water.

The muddy mugger croc

EXPLODES

from the pond

and

drags

the

deer

under.

Dinnertime.

THE ZOMBIE FISH

In the deep waters of Lake Malawi, in eastern Africa, a **CICHLID** (*si-klid*) fish seems to have gone belly-up.

It was fine just a moment ago, happily swimming past all the other fish. Then, quite suddenly, it sunk to the bottom of the lake. *Poor little fish.*

Except . . . this fish is FAKING!

It's playing dead, and with its blotchy scales, it certainly *looks* like it's rotting.

Hmm, it's a tasty option for nearby scavenging fish, so they swim in for a closer look at the corpse.

Just as they're about to take a nibble . . .

IT'S ALIVE!

The zombie cichlid comes back to life and **attacks**.

Poor little fish indeed!

INHABITANTS
OF FRESHWATER WORLDS

IT'S RAINING FROGS

Thousands of **GLIDING TREE FROGS** appear to be falling from the sky in Costa Rica, Central America. The rains have arrived, turning freshwater ponds into tadpole nurseries. It all begins when the females leap down from the rainforest's canopy to lay their eggs on leaves hanging above the pond. But as soon as a female finds the right leaf, she's mobbed by males to fertilize her eggs. They can outnumber the females nine to one. But watch your step, boys, there's danger below.

A few days after the frenzy, the tadpoles wiggle free from the eggs and PLOP! They drop into the freshwater pools – their journey has just begun.

GOBY FISH

UNLIKELY MOUNTAINEERS

Millions of sea-faring **GOBY FISH** battle upstream through river rapids on Bioko Island off Equatorial Guinea, Central Africa, to breed. But to complete the journey they must climb UP a wall found behind a fast-flowing waterfall. Some wriggle their body, while others use their mouth as a suction cup to scale the rock face. Less than 1 per cent reach the breeding ground at the top.

IS IT A BIRD? IS IT A PLANE?

Oh wait, yes, it is a bird. It's an **AFRICAN JACANA** and he's a SUPER DAD! He's so committed to his dad duties that he quite literally takes his newly hatched chicks under his wing. Look closely and you can see its little legs dangling out from under his feathers. They'll be safe there, tucked away from **NILE MONITORS** and **CROCODILES**. And that's not all, with his long legs and toes spread wide, the jacana then 'walks on water', striding across the lily pads. Is there *anything* this superbird can't do?

AFRICAN JACANA

NILE MONITOR

PAINTED WOLF

ACHOO!

A pack of **PAINTED WOLVES** has started sneezing in the Okavango Delta in Botswana, southern Africa. It's not the dust that's got up their noses, it's a vote! Sneezing is how the pack agrees whether to hunt or not, and if enough of them are in, then it's time to move. Wait for it, wait for it . . .

ah-ah-achoo!

Achoo!

And they're off, splashing through the flood plains chasing a **LECHWE** (*letch-way*), a type of water-loving antelope.

Achoo!

FIGHTING THE FLOW

Over many years, humans have come up with clever but unnatural ways to make sure we always have available freshwater.

We've *stopped* many rivers from flowing by building mighty concrete dams. We've *controlled* freshwater's path, directing it to where we need it most. And we've *taken* freshwater for our own uses – LOTS of it. But we forgot one important thing . . .

this water isn't just ours for the taking.

In Pakistan, the mighty Indus River's flow is now controlled by a complex network of dams and canals so humans can send water in the direction of thirsty crops. While this benefits humans it creates a *barrier* for wildlife.

Swimming through these muddy waters is one of the rarest of all aquatic mammals – the **INDUS RIVER DOLPHIN**. It's evolved to have eyes so small, it's essentially blind. Instead it releases sound waves that bounce off objects (echolocation) so it can find food, including shrimp and fish. But in her quest for fish, this shy dolphin accidentally swam through the open gate of a canal and is now trapped hundreds of kilometres from the main river. And water levels . . . *are dropping.*

When this happens, dolphins depend on humans to see them and save them. In the space of a month, nearly thirty dolphins were rescued and returned by truck to the Indus River. And it's these efforts that are helping to save this endangered species from the brink of extinction.

It worked this time . . . but what about next time?

Water is life for humans, but it's also life and home for millions of animals and plants. For everyone to benefit it needs to run free.

Freshwater needs to *flow*.

COASTS
WHERE WORLDS COLLIDE

Where the land meets the sea are Earth's constantly changing coastlines. They're found on every continent and in every climate, but no two stretches are ever alike.

In Africa, desert dunes dip their toes in the ocean, while at the poles, icy fingers reach across chilly seas. Yet on every coastline, opportunities for life abound. From mangrove forests and rocky clifftops, to sheltered seas and rock pools, these coastal habitats provide ideal places to shelter, feed and raise the next generation.

The day is dawning over the Sea of Cortez, off Mexico's west coast, and thousands of **MOBULA RAYS** have come to breed. They burst from the water in every direction.

Are they *swimming* or *flying*?
It seems like it's both.

Like birds, they gracefully flock underwater before using their flat, diamond-shaped bodies and huge 1-metre wingspan to perform gravity-defying acrobatics. They can leap 2 metres above the waves.

SPLASH!

The higher the males **flap**, the louder they **flop**.

Perhaps they do it to attract a mate?

MOBULA RAY

It's certainly caught something else's attention . . .

ORCA

ORCAS have detected the rays and these expert pack hunters are now working together to whip them into a frenzy. A few exhausted rays become separated from the group, and that's when the orcas strike. Maybe all that leaping *wasn't* such a good idea.

Discover more surprising behaviours from creatures living life on the edge – those places where land meets sea.

STORIES
FROM THE COASTS

PEEKABOO PUP

Until now, this **CAPE FUR SEAL** pup has lived on the edge of two worlds – the crowded cliffs above and the wild sea below.

But look at those flippers . . . there's no question where this pup *truly* belongs.

In the water, he quickly goes from awkward landlubber to sleek swimmer, twisting and twirling through the sea off South Africa's coast. But the fun and games are only just beginning. These shallows are patrolled by a **fearsome** ocean predator –

the **GREAT WHITE SHARK.**

The pup and the shark are about to play a game of peekaboo, and there's just one rule – always see the shark, but don't let the shark see you.

But the pup soon strays into open water. *Uh oh*, he's been spotted, and now this pup has found himself in a toothy game of tag.

Luckily for the pup, the adults were watching, and here comes the angry mob now. One by one, the fully grown seals take a run at the shark, harassing it from behind, before darting back to the safety of the group.

They do it over and over again, until the shark finally gets the message . . . **don't mess with the mob.**

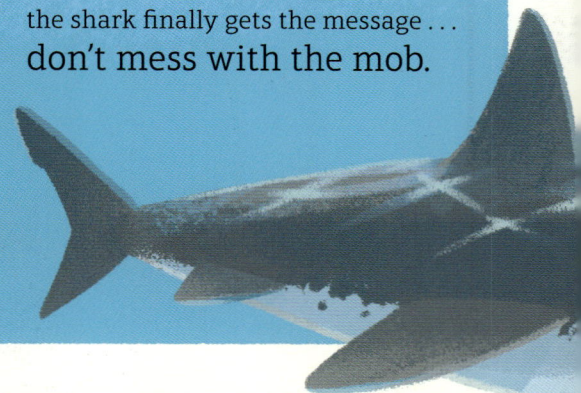

NO ANGEL

Gracefully *flying* through the chilly polar waters is one of the ocean's most mesmerizing creatures – the **SEA BUTTERFLY**, a kind of **ZOOPLANKTON**.

It flutters its wings like a butterfly but looks more like a snail. And in a tragic twist, it's a delicious meal for its equally beautiful but larger cousin, the **SEA ANGEL**.

But don't be fooled . . . this predator is *no* angel.

The sea angel closes in on the sea butterfly, one graceful glide at a time, appearing to move in s l o w m o t i o n.

But with no eyes to see, he sneaks up using sense, not sight.

He's reached top speed now – 10 centimetres a second.

This is going to take a while . . .

Closer,

c-l-o-s-e-r,

nearly there . . .

When he's almost within striking distance, the sea angel's mouth transforms into tentacles. They trap the sea butterfly and

squeeeeeeze.

Then, using tiny hooks, the sea angel scoops the butterfly from its shell.

As with the sneaky attack, the act of swallowing will be *unbearably* slow.

INHABITANTS
OF THE COASTS

SOLAR-POWERED SNAKE

A **WANDERING GARTER SNAKE** is sunbathing on a rocky beach along Canada's west coast. As the summer sun warms her up, she's ready for a spot of lunch. She tests a rock pool's waters with her tongue, sensing this is where her meal is hiding. She *sloooowly* enters the freezing water, just a few scales at a time. Then when the time is right . . . she strikes!

WANDERING GARTER SNAKE

CARIBBEAN FLAMINGO

PLEASE DON'T SPIT . . .

. . . unless you're an **ARCHERFISH**, in which case, fire at will! This mangrove-dwelling fish in Indonesia lurks near the water's surface, waiting for unsuspecting insects to come along.

ARCHERFISH

It then aims, squirts a jet of water and knocks the insect clean off a twig. BULLSEYE!

IT LOOKS LIKE RAIN

This muddy mound is in fact a **CARIBBEAN FLAMINGO** nest on the coast of northern Mexico. And cradled in a groove at the top is a single white egg. The nest stands as tall as a 30-centimetre ruler to keep the egg high and dry from dangerous flood waters that rise as storms approach.

SISTERS IN THE SAND

These **DESERT LION** sisters are all alone in a sea of sand. Orphaned as cubs, for five years they've wandered the Namib desert in south-western Africa.

Never taught to hunt, they've scavenged on what little they can find. Tired and terribly hungry, their search for food has led them someplace no lion has ventured for almost forty years . . . the coast.

This is the Skeleton Coast, a long-forgotten hunting ground for these lions, one that's full of strange new sights and smells. And so, these nervous sisters wait in the sand dunes, finally plucking up the courage to emerge under the cover of night. And when they do, they discover the beach is *alive* with opportunities.

There are **FLAMINGOS**, **CORMORANTS** and even a **FUR SEAL** in the surf. There's only one problem . . . *rarrrrrrr* . . . like their cat cousins, these lions aren't always keen on getting their paws wet!

Twenty-five years ago, around twenty desert lions remained in the wild due to conflict with humans. But through careful conservation efforts, they've made an extraordinary recovery, with their numbers slowly rising to over a hundred.

They now take brave steps forward into a new world, providing fresh hope for the future of their species.

EXTREMES

BORN SURVIVORS

There are extreme places on Earth, where every day is a struggle to survive. From blistering deserts and icy tundra, to towering mountains and caves stretching deep underground, they are some of the most hostile and dangerous corners of our planet.

Yet over millions of years, some creatures have evolved to survive in these places where others cannot. Their adaptations make them just as powerful and beautiful as the extreme habitats they call home.

STALAGMITES

BLIND CAVE FISH

TERRESTRIAL CRAB

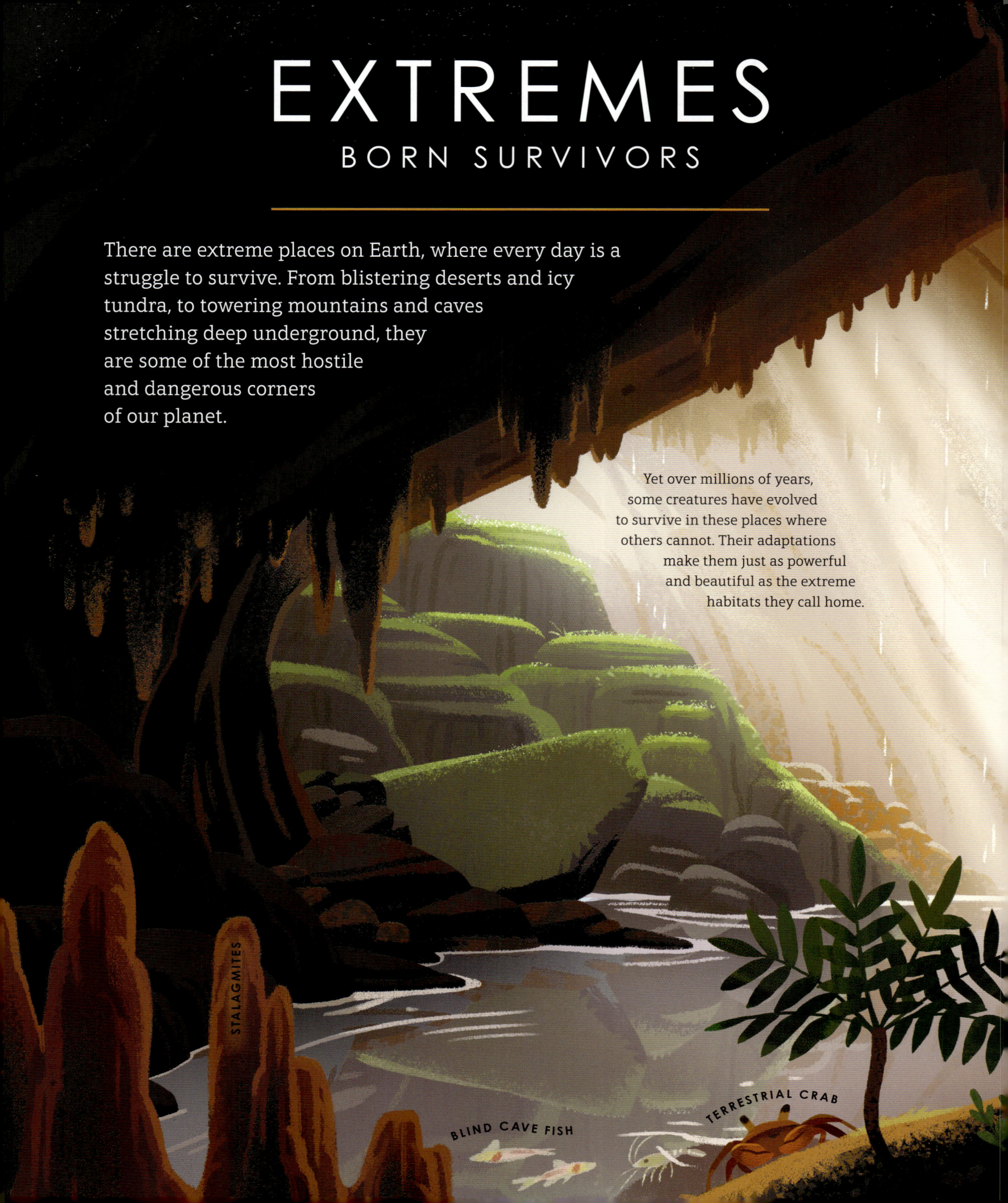

Hidden under the Vietnamese jungle in Southeast Asia is a place of light and dark –
Hang Son Doong, the world's largest-known cave passage. It was accidentally
discovered in 1990 by a local villager. And almost everything that lives in the cave
has flown, fallen or flowed through holes in the ground, to live inside Earth's belly.

Over millions of years, a fast-flowing river has worn away the limestone
to form cave chambers – one is even large enough to fit a jumbo jet inside.
In places where the roof has collapsed, light filters through the gloom,
allowing an underground forest to grow. **BIRDS**, **BATS**, **SQUIRRELS**,
SNAKES and **GECKOS** can all be found here.

STALACTITES

SCUTIGERA

Listen closely . . . the constant *drip, drip, drip* of
water leaves behind minerals that form giant
STALAGMITES and **STALACTITES**, as well as cave
pearls that look like beautiful white marbles lining
the floor.

WHITE
WOODLICE

Travel further into the cave, and the light begins to
fade, until darkness takes hold. Here in the pitch-black,
creatures have unexpectedly evolved to survive without
light, from **WHITE WOODLICE** to **BLIND CAVE FISH**.

CAVE-
DWELLING
SPIDER

CENTIPEDE

Discover just how far life
will go to survive in more
of Earth's extreme habitats.

STORIES
FROM THE EXTREMES

HOP TO IT!

It's early June and summer is dawning some 2,000 metres up in the French Alps. The sun is high, the snow is melting . . . and *someone* has overslept.

There's a rustling and a shuffling under the snow. Then . . .

POP!

A **FROG** emerges from his snowy hideout.

He's LATE. He's also a bit stiff. You would be too if you'd been hibernating all winter long beneath the snow. But there's no time to waste; he needs to find a mate before the fleeting warmth passes and the pond refreezes. The other frogs will already be gathering in an ice-free pond somewhere on these mountains, so that the males and females can pair off.

This little frog – no longer than a crayon – now has a very BIG and risky journey ahead.

It's time to make a start – this brown blob hops and flops across the stark white snow, but there are dangers lurking all around.

A **FOX** sniffs the air.

As the frog hops on, a
shadow circles above.

Still the frog hops on.

Then, in a stroke of luck, he tumbles into a tunnel.

It twists and turns, until *finally*
the frog arrives at a pond.

Oh dear, it's like frog soup in here, with the females
outnumbered three to one. But in all the frenzy to find a mate *in* the
pond, the other frogs haven't noticed a lone female *out* of the water.

The frog swims to the edge and calls
out to her with his most fetching C R O A K !

And with that, the match is made.

Better late than never.

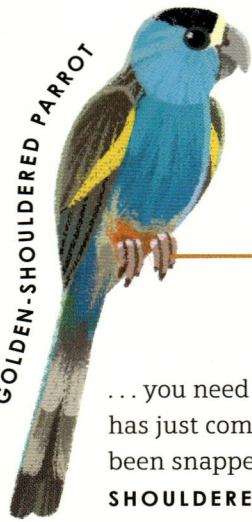

INHABITANTS
OF THE EXTREMES

IN THE HOT AUSSIE OUTBACK . . .

. . . you need a *cool* home. Termite Tower has just come on the market, and it's been snapped up by a pair of **GOLDEN-SHOULDERED PARROTS**. They swiftly move in, tunnelling out a nest to raise their young. Termite Tower has loads going for it. It's sturdy – made from soil, termite spit and poo – so it should keep the weather out. And it's air-conditioned too, with plenty of ventilation.

In fact, it's such *hot* property, it was still standing even after a fire destroyed everything around it. As for the chicks . . . they were safely tucked inside. *CHEEP CHEEP!*

GOANNA

BLAZING BLUE

At night, the **KAWAH IJEN** *(ka-wah ee-jen)* volcano in Indonesia, Southeast Asia, appears to glow with *blue* lava. This volcano releases incredibly high levels of sulphuric gases that burn blue and bright when exposed to oxygen in the air and the volcano's extreme temperatures.

ARCTIC WOLF

WANDERING WOLVES

On the remote Ellesmere Island, off the coast of Greenland, **ARCTIC WOLVES** wander across the tundra – the treeless plains, where much of the ground has been permanently frozen for thousands of years. These white wolves patrol this vast territory, hunting down any prey that passes into it, including much larger **MUSKOXEN**, **CARIBOU** and **ARCTIC HARES**.

LIFE ON A LONELY MOUNTAIN

Rocky mountains rise above Mongolia's Gobi desert in Central Asia.

Temperatures here can swing from a searing 35 degrees Celsius in summer to a freezing -35 degrees Celsius in winter. It would be easy to think this barren and rugged environment is empty of life. But *something* is prowling high up on the ridge, its grey coat making it all but disappear against the cliffs. It's a **SNOW LEOPARD** – perfectly adapted to life on the edge.

Of the six thousand or so snow leopards that roam the planet, almost a thousand are found here in Mongolia's mountains. But their homes in this region were under threat from humans mining for valuable metals and fossil fuels. Fortunately, conservationists won the battle to turn the area into a nature reserve, giving mothers like this one a chance to raise a new generation of born survivors.

It's a beautiful reminder that however hot, cold, deep, high or dangerous Earth may be, something, somewhere calls that hostile place *home*.

OCEANS

WORLDS BENEATH THE WAVES

The ocean is vast and still largely unexplored, hiding worlds we're only now beginning to understand. From bustling reef cities and towering kelp forests stretching three storeys high, to seemingly endless desert plains, these habitats hide an unimaginable diversity of life. When humans journey into these worlds, extraordinary discoveries of unknown species and unseen spectacles are made.

GIANT CUSKEEL

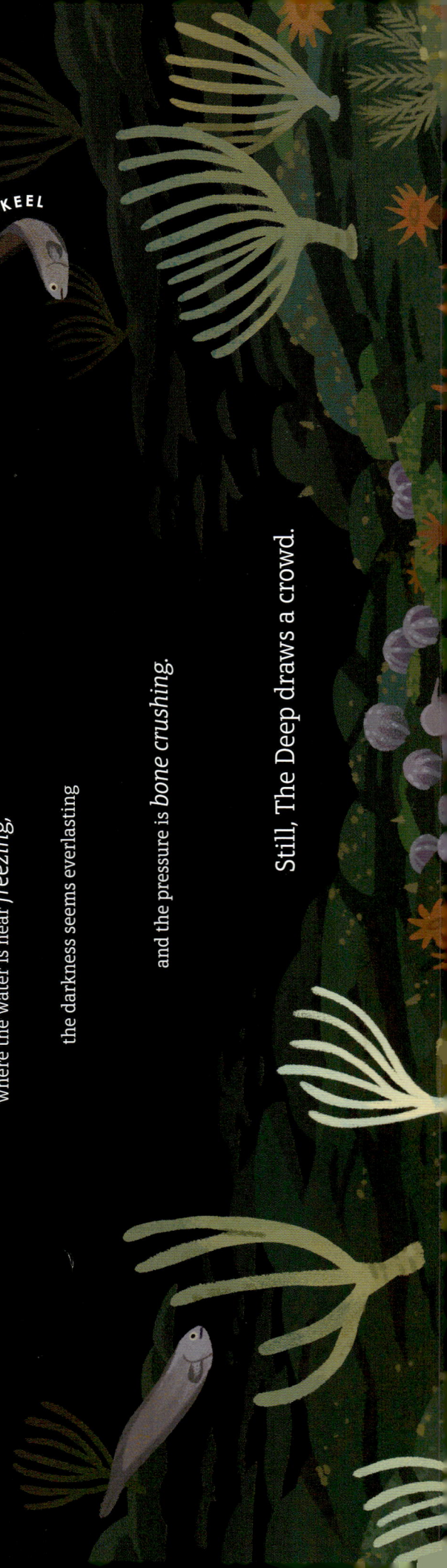

Venture deeper into the ocean . . . .

where the water is near *freezing*,

the darkness seems everlasting

and the pressure is *bone crushing*.

Still, The Deep draws a crowd.

Around 3,000 metres down, thousands of **DEEP-SEA OCTOPUSES** huddle around cracks in the seabed, called hydrothermal springs, where warm water escapes. But they're not here for a spa day . . . it's an octopus nursery, where devoted mums have come to raise their young.

Temperatures can reach 10 degrees Celsius around the springs, helping the eggs to develop. But it still takes up to two years, making it one of the longest brooding periods in the animal kingdom, and a long time to watch an egg. And watch these mums must, because creatures are waiting for the eggs to hatch so they can strike.

Two years pass in a blur of

egg stroking

and nest guarding.

DEEP-SEA OCTOPUS

With so much to do, it's no wonder that in all this time, Mum hasn't eaten a single bite. She grows weak and her eyes begin to cloud over. Hang in there, the hatchlings are coming.

Mum fends off the **SHRIMP** one last time, and as the eggs finally hatch, she quietly fades away.

SHRIMP

Journey from the deep up to the shallows and discover the surprising inhabitants hiding in worlds beneath the waves.

She's put *everything* into raising her young, and as is the way of most octopus mothers, she's made the ultimate sacrifice. It's a fascinating behaviour, and with many species still undiscovered, it's a reminder that we have so much to learn.

ANEMONE

INHABITANTS
OF THE OCEANS

A MIDNIGHT FEAST

Somewhere on the sandy seafloor, an **ANGEL SHARK** is hiding. It has a wide, flat body, helping it almost vanish in this giant kelp forest. Now, after days of staying still, the wait is nearly over.

It's coming up to midnight as a baby **HORN SHARK** slowly swims by. It gets its name from the two horn-like ridges behind its eyes. When it's close enough, the angel shark attacks with lightning speed, swallowing the horn shark whole. But this feast is *far* from over. Horn sharks have two dorsal fins with sharp spines, and they make for a *very* prickly mouthful, forcing the angel shark to spit it out. The lucky little horn shark will live to see morning.

ANGEL SHARK

HORN SHARK

MEET WARTFACE . . .

. . . isn't he magnificent? He's a **WARTY FROGFISH** that lives on coral reefs throughout the tropics. He's full of tricks. As well as swimming, he can 'walk' under water by using his leg-like fins to move along the ocean floor. And for his next trick, he's going to disappear. Now you see him . . . now you don't!

That's because Wartface looks like a sponge or a lump of coral, which helps camouflage him against the reef. And in his last amazing illusion, Wartface dangles a fishing-rod-like lure between his eyes. It looks like a shrimp, tempting fish to take a closer look. SNAP! The show's over.

WARTY FROGFISH

ATTENNNNN-TION!

This is **SERGEANT MAJOR DAMSELFISH** reporting for duty. Over the next few weeks, he'll be:

SERGEANT MAJOR DAMSELFISH

HERMIT CRAB

A builder. Clearing away seaweed and rubble to make the perfect nest.

A dancer. Bopping away to catch the eye of passing females so they'll lay their eggs in his nest.

A guard. Working day and night for a week to patrol thousands of his eggs, dragging off imposters, including **HERMIT CRABS**!

And importantly, he does it all as a single dad. Without him, these hatchlings wouldn't have made it. Great job, Dad!

ALIENS OF THE DEEP

Astonishingly, in the deep sea, there is more life than anywhere else on Earth – with countless species still undiscovered.

Floating in the Twilight Zone, between 200 and 1,000 metres down, is the newly discovered **GLASS SQUID**. It's completely see-through!

The **GULPER EEL** moves even deeper through the pitch-black Midnight Zone, 1,000 metres below the ocean surface. It scoops up water into its pelican-like mouth, swallowing **CRUSTACEANS**, **FISH** and **SQUID** whole.

GLASS SQUID

GULPER EEL

THE
CASTAWAY CRAB

On the open ocean, floating seaweed becomes a *lifeline* –
a welcome shelter for young castaways.

It can give turtle hatchlings and fish somewhere to feed and provide cover
from the big blue's mighty predators. But something else now bobs in the
waves. Plastic. It never really disappears, it just weathers away into tiny pieces
that animals mistake for food or get tangled up in. And one of the deadliest types
are ghost nets. These discarded fishing nets are almost invisible to wildlife and
are a growing problem that conservationists are working to solve.

As destructive as ghost nets are, one has unexpectedly become a raft.

This **COLUMBUS CRAB** has been clinging on to a ghost net since he was young,
having drifted there on the ocean current. He's watched the world wash by, but he's
now grown up and ready to explore the big blue. The trouble is, he's not much of a
swimmer. And that makes things a *bit* tricky, because as far as his beady eyes can
see . . . there's water.

But wait . . . a bubble. And another one. Then along swims a **LOGGERHEAD TURTLE**.

This is Columbus's big moment – it's time to hitch a ride. He bends his legs, and brave little Columbus *finally* takes the plunge.

Oh dear, is he swimming or sinking? C'MON, COLUMBUS – YOU CAN DO IT!

But with an outstretched claw . . . Columbus finally latches on.

His new mobile home is very nice. The turtle's shell is covered in **BARNACLES**, perfect for Columbus to eat. And there's a cosy nook to live in under the shell too. And that's when Columbus's day goes from good to great. Also hiding under the shell . . . IS A MATE!

It's a small yet beautiful example of the interconnectivity of all life in the ocean. That even in the blue wilderness, a lonely crab can not only find a lift – but love.

HUMAN WORLD
NATURE'S NEW WILD

For thousands of years **HUMANS** have made their way into every corner of the planet – eight billion of us now make our mark on the world.

And wherever we go, nature has had to adapt – to either put up with us . . . or take us on. So, when an opportunity arises in our built-up world, many animals seize it. Wildlife is now fighting for space in our houses, on our farms and in our cities.

If you look close enough,
armies of them are
marching
right
under
our
feet.

PIGEON

HUMAN

PAVEMENT ANT

Welcome to New York City, where for every person there are two thousand ants – most of them **PAVEMENT ANTS**. Like early New Yorkers, they arrived from Europe inside ships, probably around two hundred years ago, and are now right at home on the busy city streets. To avoid being trampled, they scurry along cracks in the pavement. And just like the locals, they love fast food.

Get it
while it's
HOT!

These pavement ants have become hot-dog enthusiasts. New Yorkers throw away a million tonnes of food every year. And here, under the bright lights of Broadway, the ants get to work chowing down on the scraps. They eat the equivalent of about 60,000 hot dogs annually – helping rid the streets of leftovers.

Discover some of the other fascinating ways nature is not only keeping pace in our human world, but becoming bold enough to take us on.

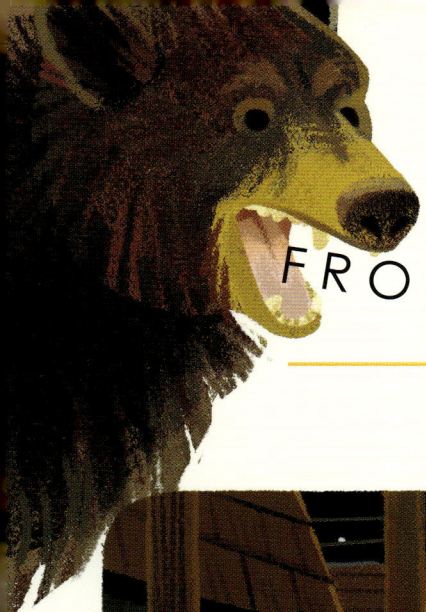

STORIES

FROM THE HUMAN WORLD

DID SOMEBODY SAY PIZZA?

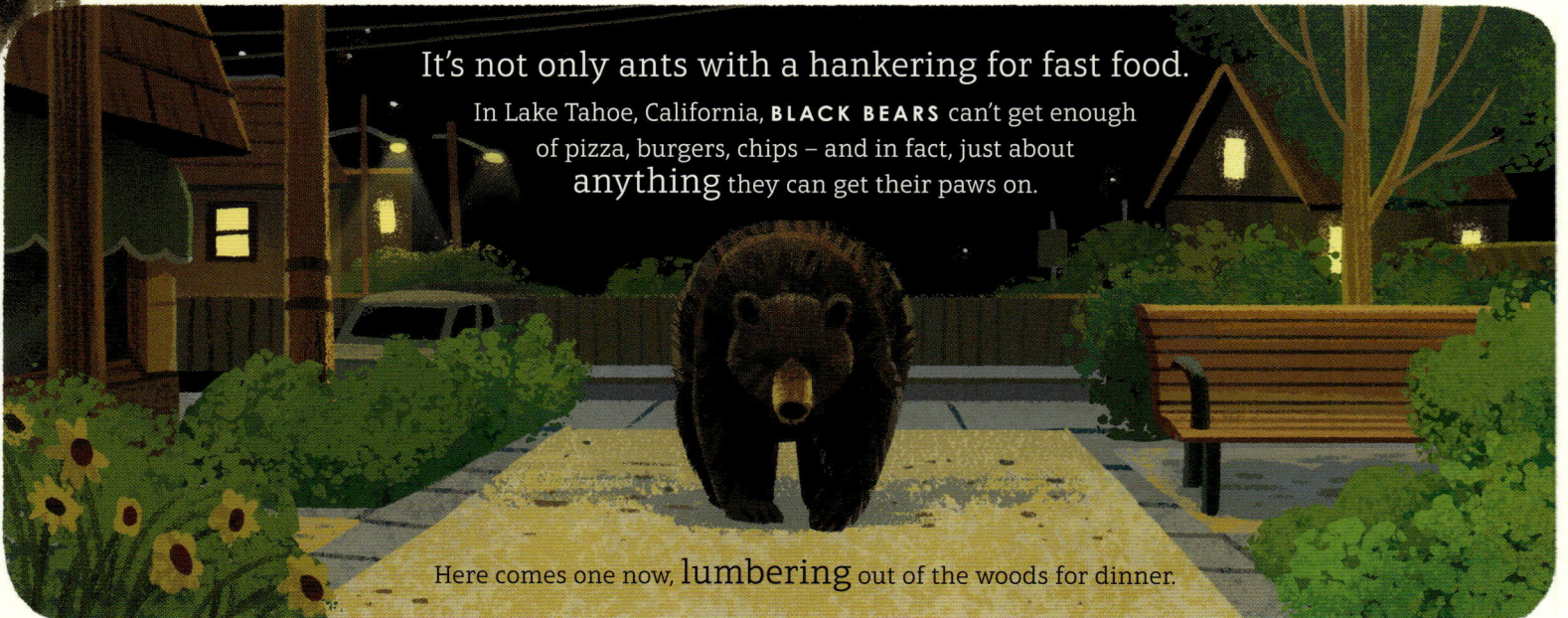

It's not only ants with a hankering for fast food.
In Lake Tahoe, California, **BLACK BEARS** can't get enough of pizza, burgers, chips – and in fact, just about **anything** they can get their paws on.

Here comes one now, **lumbering** out of the woods for dinner.

He sizes up a rubbish bin, and OOOOOOF, he's in.

This furry felon has been spotted eating leftovers in homes all over the area. The locals are 'bear aware' and know not to leave food about. But nobody expected one to pop to the shops instead.

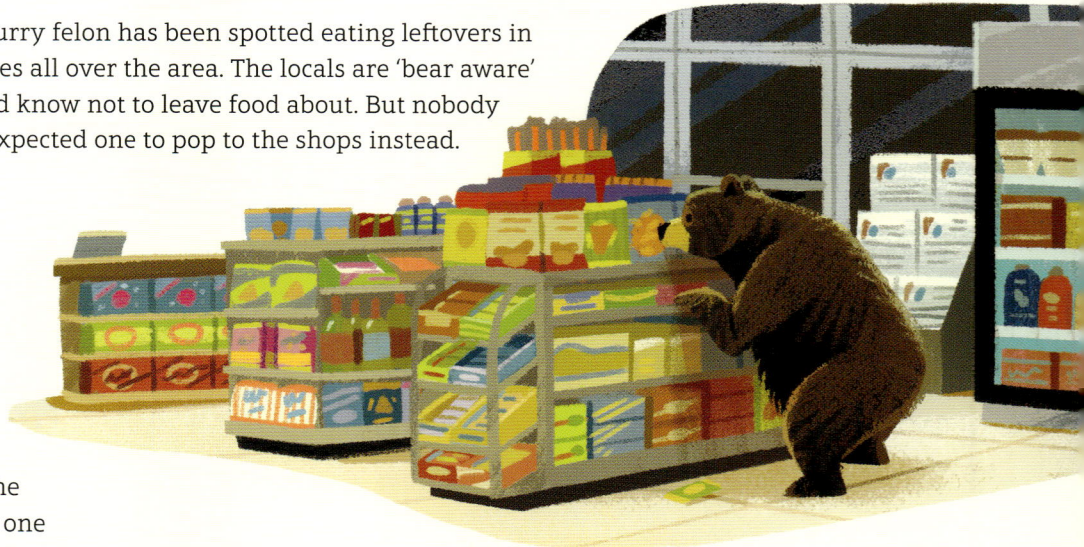

With a full belly, it's time to head back into the woods. With all that fast food, bears like this one are fast expanding. At over 200 kilograms, he's about twice the weight of the average black bear.

Black bears in this area have discovered the food in the shops and the bins is much richer than what's found in the woods. Perhaps they *won't* hibernate this year.

After all, why sleep when you can snack?

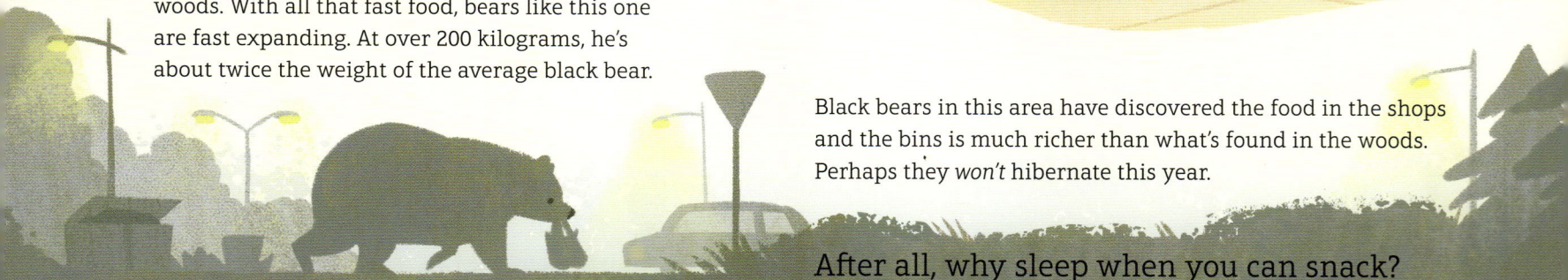

CALM COBRAS

The moment deadly **MONOCLED COBRAS** slither into one of four villages in West Bengal, India, they're no longer scary – they're sacred.

The villagers believe these cobras are **spiritual** creatures, and so they show them **respect**.

Free to swim in the ponds where women wash their pots or slither past children's toes in the schoolyard.

And the cobras are not just in their houses . . . they're sometimes in their beds!

For many people this may seem a bit too close for comfort. After all, snakebites kill around 58,000 people in India every year.

But somehow these monocled cobras are *calmer* – they move more slowly and are less likely to bite than others. And when bites *do* happen, they're often 'dry bites' – painful but lacking venom.

It's an incredible example of how when humans make space for animals — even dangerous ones – we can both live calmly, *side by side.*

INHABITANTS
OF THE HUMAN WORLD

More and more, the thin line between the human world and the wild is being blurred. Are animals coming into our world, or are we expanding into theirs?

HITTING THE TOWN

An **INDIAN RHINOCEROS** is spotted strolling through a small town in Nepal at night. He wanders out of his forest and down the street to reach grazing grounds on the other side of town – paying no attention to the humans stopping for selfies. People have hunted the Indian rhinoceros almost to extinction. But through careful conservation, numbers have increased to over 3,500 in India and Nepal.

INDIAN RHINOCEROS

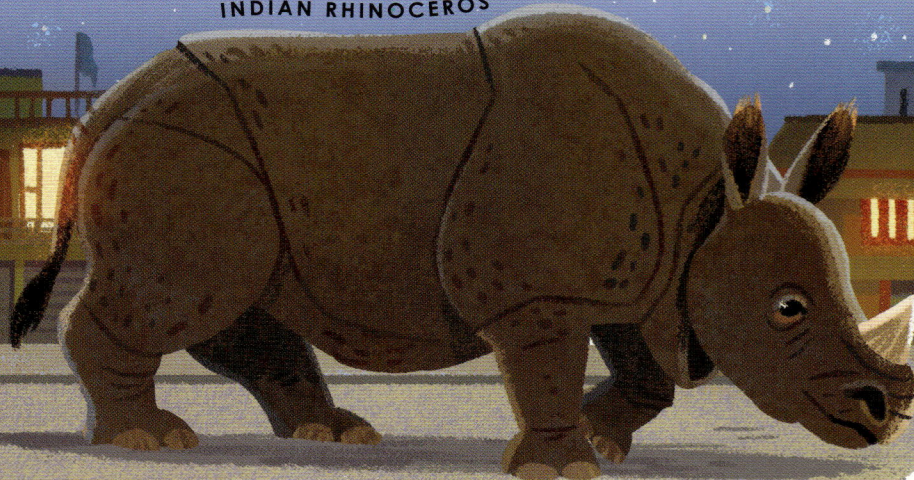

DESERT LOCUST

FREEZE, FROGMOUTH!

Don't move. Don't even blink. There's a **CAT**! Although typically found in Australia's forests, **TAWNY FROGMOUTHS** have become city dwellers. The bright streetlights help these nocturnal birds to hunt **MOTHS**. There's just one problem . . . feisty felines. When they see a neighbourhood cat, they must stay statue-still to avoid being spotted.

TAWNY FROGMOUTH

CAT

EATING *EVERYTHING*

As their name suggests, **DESERT LOCUSTS** can be found in dry habitats across Africa and Asia. But scientists believe that as climate change causes unseasonal high levels of rainfall in some areas, it's creating locust 'super swarms'. In the space of a day, a single swarm containing billions of locusts can take to the sky, covering hundreds of kilometres on the wind, devouring as much food as 3.5 million people could eat.

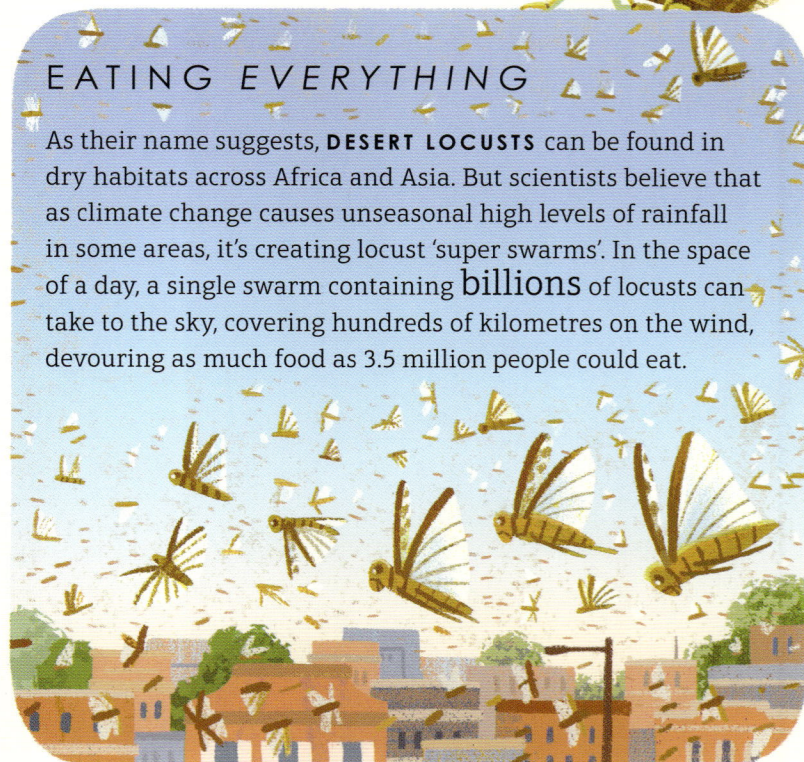

THE POWER OF POO

Thirty years ago, **HUMPBACK WHALES** had all but vanished from the waters north-east of Vancouver Island, off the west coast of Canada, because of hunting.

Overfishing had caused the numbers of their preferred prey – a small fish known as a **HERRING** – to dramatically reduce too. But the humpbacks' story was far from over.

Commercial whaling was banned in 1986, and after being left alone for many years, the humpbacks' numbers slowly began to rise. And as they returned to the waters north-east of Vancouver Island, life followed in their wake.

Humpbacks release *huge* iron-rich poos at the ocean's surface. Scientists believe these poos help feed marine plants known as **PHYTOPLANKTON**, which are then eaten by herring. And where there's lots of herrings, hungry birds follow, diving into the ocean and whipping the herring into a frenzy. The panicked fish form a tightly packed 'baitball', attracting **SEALS**, **DOLPHINS** and **SEA LIONS**, before becoming the ultimate mouthful for humpbacks as they *lunge* from below.

Today around a hundred humpbacks can be seen off the north-east coast of Vancouver Island, lunging at the herring.

And all it took was a simple change from humans to help give nature room to thrive.

WE CAN BE HEROES

Often the story of human life on Earth seems like a tragedy, where we play the villains. But that's not the full story. All around the world HEROES, just like you, are stepping up for nature. They're smart, resourceful, courageous, and if ever we needed them, it's NOW!

HUNTING CRIMINALS, NOT ANIMALS

Conservationist Trang Nguyen is risking everything to save the critically endangered **FOREST ELEPHANT**. Shockingly, forest elephant numbers have dropped by two-thirds in just nine years. Throughout West and Central Africa, forest elephants are killed for their tusks, which are sold and illegally taken out of the country. To catch these criminals, Trang and her team set a trap; she goes undercover pretending to buy the ivory. Then when the 'gang' reveal their haul, the police pounce. Every week, criminals are stopped in their tracks thanks to brave people such as Trang.

SAVING SANTIAGO

Santiago, a type of **HARLEQUIN FROG**, would have been one of the last of his kind were it not for scientist and explorer Jaime Culebras and his team. Like so many amphibians, Santiago's species is facing extinction as they die from the chytrid (kit-rid) fungus. Determined to find him a mate to continue the species, Jaime set off on foot into the remote cloud forests of Ecuador, South America. Night after night, he searched high and low, but the streams that should have been filled with frogs were empty. Then, one night, drenched and exhausted, Jaime found a female hiding in the bushes. Luckily for the harlequin frogs, Jaime has since found more of their kind. Little frog families are now coming along nicely in a breeding facility, providing hope for threatened species everywhere.

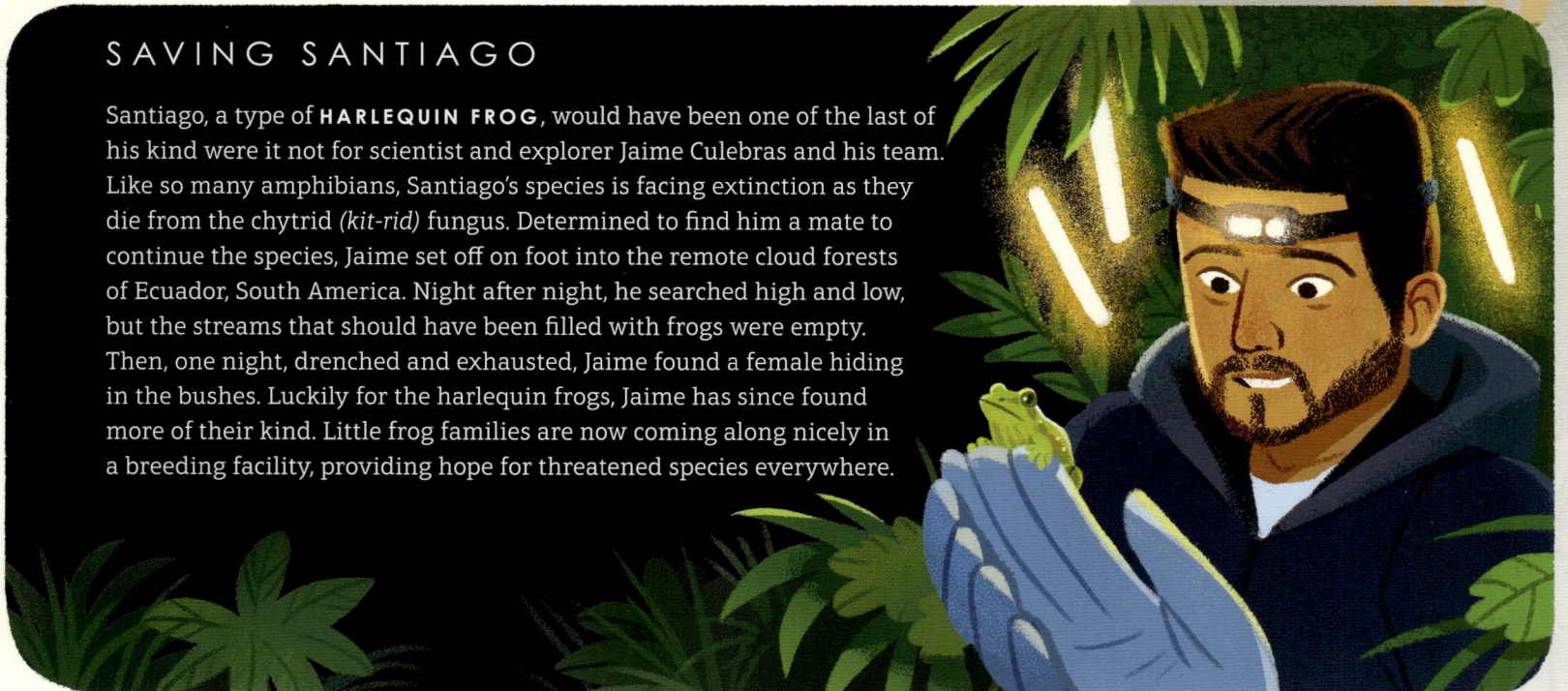

NOT ALL HEROES WEAR A CAPE . . .

. . . in the battle against climate change – some wear suits and their weapons are words. This is Mohamed Nasheed; he was once president of the island nation of the Maldives in the Indian Ocean. It's the world's lowest-lying country, sitting barely 2 metres above sea level. If Earth gets much hotter, melting sea ice and glaciers will cause sea levels to rise, putting the Maldives (and similar communities) at risk of being swallowed by the ocean. Knowing this, Mohamed shows up at climate conferences, using his powers of persuasion to try and convince politicians to take actions that will limit global warming to 1.5 degrees Celsius and protect his island home. He'll never stop trying, but will it be enough?

SPEAKING UP FOR THE FOREST

The Amazon is home to more species of plants and animals than any other land-based habitat on Earth and more than a million Indigenous Peoples too. For generations they've lived in harmony with nature, but that's fast changing as their forest home is illegally dug up by miners for oil and gas, and cut down by loggers for trees and cattle ranches.

Rather than raising bows and arrows to defend the forest and themselves as they once did, many Indigenous communities are raising cameras to share these illegal activities on social media for the world to see.

This is exactly what the women of the Munduruku tribe are doing. And their leader, Alessandra Korap, took their message even further . . . to Brazil's capital, Brasília. Here she joined thousands of other Indigenous Peoples in a ten-day protest to speak out for Indigenous Peoples' rights and the Amazon rainforest.

EVERYDAY
HEROES

Everything in nature is connected in a marvellous and messy web of life.
So, when we take something from Earth, something else feels that loss,
from the smallest bug to the biggest beast.

This is *their* planet as well as ours.

That's why we need to treasure everything Earth shares with us.
Remember, drinking water is pulled from lakes and rivers,
food is grown where forests and grasslands once grew,
and a scrap of paper once stood proud as a tree.

Here's how you can be an everyday hero and
take care of Earth's precious resources too:

SAVOUR EVERY CRUMB

Around 40 per cent of every beast reared or crop grown each year is *not* eaten. But if
we all shopped smarter and ate up our leftovers, together we could save huge patches
of wilderness from being destroyed.

PICK UP THE PIECES

When you head out for a walk, why not help leave nature better than you found it by
picking up any litter you spot?

BE WATER-WISE

Turning off the tap when you clean your teeth or wash your hands can save 6 litres of
water a minute – that's enough to fill half of an average household bucket. Now imagine
if everyone in your class, school or town did that . . . that's a LOT of buckets.

PROTECT YOUR PAPER

Every year roughly three billion trees are cut down to make paper packaging. That's enough
trees to cover all of Germany. We can all help cut down on paper instead of trees by refusing
single-use paper products, reusing wrapping paper and recycling paper packaging.

Humans have walked this planet for over 300,000 years, and yet we're still finding our place
in it. We're seeing the life-giving power of water when it's set free to flow across the land.
We're discovering that even species pushed to the brink of extinction can make extraordinary
recoveries. And we're realizing that Earth's most extreme reaches still have secrets to share.

But most important of all, we're learning that when we
tread lightly on the world and take only what we need,
all of life on this incredible planet can thrive.

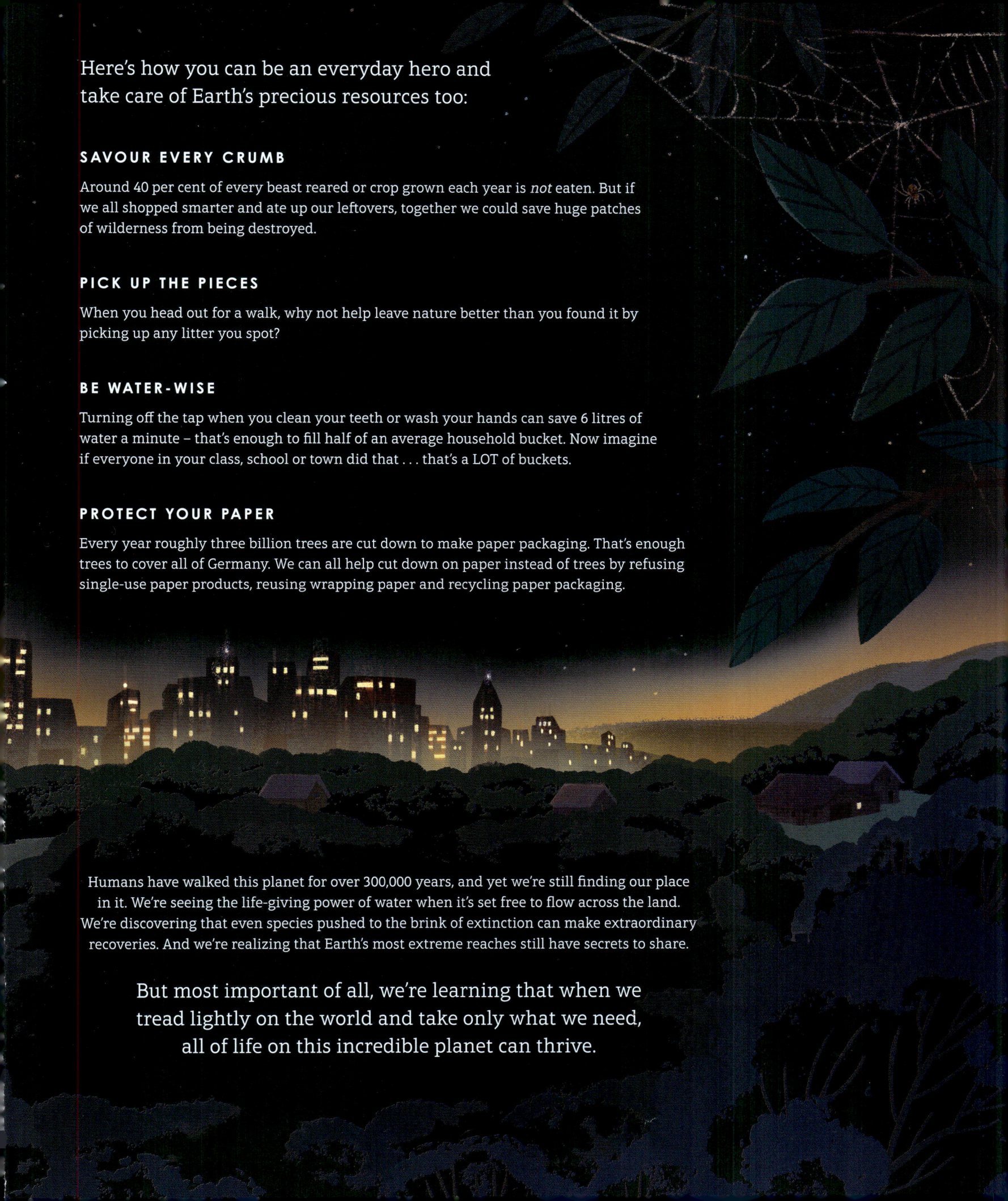

BEDROCK
Solid rock found under the soil.

BIODIVERSITY
A huge variety of life in one place on Earth – from bacteria and plants to fungi and animals.

BIOME
A large area of Earth with a particular soil, climate and wildlife that lives there. Major biomes include forests, grasslands, deserts, tundras and aquatic (freshwater and marine) habitats.

CANOPY
The dense and crowded uppermost layer of the forest.

CARBON DIOXIDE (CO₂)
An odourless and colourless gas that occurs in Earth's atmosphere, created when humans (and other animals) breathe out, or emitted as pollution from factories and vehicles.

CENOTE
A hole in the ground that forms when the limestone bedrock naturally collapses to reveal a freshwater pool underground.

CLIMATE CHANGE
Long-term changes in Earth's atmosphere and average temperatures.

DECOMPOSING
The process of decay, when something breaks down into smaller parts.

DESERT
Covering around 20 per cent of Earth's land, these habitats receive very low levels of rainfall, typically around 25 centimetres a year.

ECHOLOCATION
When an animal works out the location of something by releasing sound waves that bounce off the object.

EPIPHYTE
Plants that grow above the ground using other plants or rocks for support.

GRASSLAND
Covering around 40 per cent of Earth's land, these habitats are dominated by grasses and grass-like plants rather than large shrubs or trees.

HABITAT
A plant, animal or other organism's natural home in the wild: from a woodland to a coral reef.

LAVA
Molten rock that comes out of the Earth at very high temperatures.

MIGRATING

When, in a certain season, an animal travels from one habitat or region to another.

PHYTOPLANKTON

Microscopic, plant-like organisms that drift with the ocean current – the foundation of many marine food chains.

SAVANNAH

An open grassland, often dotted with trees or shrubs, found between tropical rainforests and deserts.

STALACTITES

Rock formations shaped like icicles that hang from cave ceilings.

STALAGMITES

Rocky mounds that grow upwards from cave floors.

STEPPE

A grassy, treeless plain found in dry climates between tropical and polar regions.

TEMPERATE RAINFOREST

A rainforest that grows in milder climates outside of the tropics and experiences lots of rainfall but cooler temperatures.

TROPICAL RAINFOREST

A warm forest habitat found near the equator, between the Tropics of Cancer and Tropics of Capricorn, that receives more than 2 metres of rainfall a year.

TROPICS

The regions of Earth that cover the middle of the globe, just to the north and south of the equator.

TUNDRA

Found in the Arctic and high mountain areas, this is a flat and treeless area where the ground is permanently frozen.

ZOOPLANKTON

Small and weakly swimming animals that drift on ocean currents, such as krill, larvae and jellyfish.